Preface

As a science enthusiast and space lover, I have always been fascinated by the mysteries of the universe. The solar system, with its countless planets, moons, asteroids, and comets, has always held a special place in my heart. I have spent countless hours researching and learning about our neighboring planets, and I'm excited to share my knowledge with you in this book.

In these pages, you'll find a comprehensive guide to our solar system. From the rocky, inner planets to the gaseous giants in the outer reaches, each planet and moon is explored in detail. You'll learn about the unique features and characteristics of each world, including its atmosphere, surface features, and any moons it may have.

But this book is more than just a scientific guide. It is also a celebration of the wonders of the universe and the human spirit of exploration. Throughout history, humans have looked up at the stars with a sense of wonder and curiosity, and have dared to dream of exploring the depths of space. I hope this book will inspire you to continue that tradition and to always look up at the sky with a sense of wonder and awe.

Finally, I want to acknowledge the many scientists, researchers, and explorers who have dedicated their lives to understanding the universe and our place within it. Without their hard work and dedication, this book would not have been possible.

Thank you for joining me on this journey through our solar system, and I hope you find this book both informative and inspiring.

Author
Abhishek Sharma

Chapter-1

Solar System

The solar system is a collection of celestial bodies that are gravitationally bound to the Sun, a star located at the center of the system. The solar system includes eight planets *(in order from the sun: Mercury, Venus, Earth, Mars, Jupiter, Saturn, Uranus, and Neptune)*, five dwarf planets *(Ceres, Pluto, Haumea, Makemake, and Eris)*, many moons, asteroids, comets, and other small bodies.

The solar system formed about *4.6 billion* years ago from a giant cloud of gas and dust, known as the solar nebula. The nebula began to collapse due to its gravity, forming a spinning disk of material around a protostar that eventually became the Sun. As the disk cooled, solid particles began to stick together, forming planetesimals, which in turn collided and merged to form the planets.

The inner solar system (Mercury, Venus, Earth, and Mars) is rocky and relatively small, while the outer solar system (Jupiter, Saturn, Uranus, and

Neptune) is composed mostly of gas and ice giants. The solar system is also home to many moons, including Earth's moon, which is the fifth largest moon in the solar system.

The solar system is constantly changing, with asteroids and comets occasionally colliding with planets or each other, and the orbits of the planets shifting over time. Scientists continue to study the solar system and its many mysteries, including the origins of life on Earth and the potential for life elsewhere in the universe.

Here are some additional facts about the solar system:

- The most massive object in the solar system is the Sun, which accounts for 99.86% of the total mass.
- The largest planet in the solar system is Jupiter, which is more than twice as massive as all the other planets combined.
- The smallest planet in the solar system is Mercury, which is only slightly larger than Earth's moon.
- The asteroid belt, located between Mars and Jupiter, is a region where many small, rocky bodies orbit the Sun. The largest asteroid in the belt is Ceres,

which is also classified as a dwarf planet.

- The Kuiper Belt is a region beyond Neptune where many small icy bodies, including Pluto, orbit the Sun.

  Halley's Comet, one of the most famous comets, orbits the Sun once every 76 years and was last seen from Earth in 1986.

- The Oort Cloud is a hypothesized region of icy bodies that surrounds the solar system at a distance of up to 100,000 astronomical units (AU) from the Sun. An astronomical unit is the average distance between the Earth and the Sun, which is about 93 million miles (149.6 million kilometers).

- The Voyager 1 and Voyager 2 spacecraft, launched in 1977, are the farthest man-made objects from Earth and are currently traveling through interstellar space.
- The New Horizons spacecraft flew by Pluto in 2015, providing the first close-up images of the dwarf planet.
- The solar system is believed to be about 30,000 light-years from the center of the Milky Way galaxy and takes about 230

million years to complete one orbit around the galaxy.

1. *Mercury*: Mercury is the smallest planet in the Solar System and is the closest planet to the Sun. It has a rocky surface that is heavily cratered and lacks any atmosphere. Its surface temperature can reach over 400 degrees Celsius (800 degrees Fahrenheit) during the day and can drop to -170 degrees Celsius (-274 degrees Fahrenheit) at night.

2. *Venus*: Venus is the second planet from the Sun and is

similar in size and composition to Earth. It has a thick atmosphere that traps heat and makes its surface temperature the hottest of any planet in the Solar System, reaching up to 470 degrees Celsius (878 degrees Fahrenheit). Its surface is also rocky, with mountains, valleys, and volcanoes.

3. *Earth*: Earth is the third planet from the Sun and is the only planet known to support life. It has a diverse ecosystem and a stable climate, thanks to its protective atmosphere, magnetic field, and distance

from the Sun. Earth has a solid surface with oceans, continents, and a variety of landforms.

4. *Mars*: Mars is the fourth planet from the Sun and is known as the "Red Planet" due to its reddish appearance in the night sky. It has a thin atmosphere and a cold, dry surface that is covered in rust-colored rocks and dust. Mars is also home to the largest volcano and the deepest canyon in the Solar System.

5. *Jupiter*: Jupiter is the largest planet in the Solar System and is fifth from the Sun. It is a gas

giant with no solid surface, but rather a thick atmosphere made mostly of hydrogen and helium. It has a complex system of rings and over 70 known moons, including the four largest moons known as the Galilean moons.

6. *Saturn*: Saturn is the sixth planet from the Sun and is another gas giant with a complex system of rings. It is similar in composition to Jupiter, but its rings make it one of the most recognizable planets in the Solar System. Saturn has over 80 known moons, with its largest moon

Titan being the only known moon with a thick atmosphere.

7. *Uranus*: Uranus is the seventh planet from the Sun and is an ice giant with a gaseous atmosphere and a solid, icy core. Its axis is tilted at an extreme angle, causing it to rotate on its side. Uranus has a system of rings and over 20 known moons.

8. *Neptune*: Neptune is the eighth and farthest planet from the Sun, and is also an ice giant with a gaseous atmosphere and a solid, icy core. It has a faint system of rings and over 14

known moons, including Triton, which is the coldest known object in the Solar System.

Sun

The Sun is the star at the center of the Solar System, and it is by far the most massive object in our Solar System, comprising more than 99% of the total mass of the Solar System. It is a giant ball of hot, glowing gas, with a diameter of about 1.39 million kilometers (864,938 miles), which is about 109 times the diameter of the Earth.

The Sun is composed mainly of hydrogen gas, which is fused in its core to form helium through a process called nuclear fusion. This process releases an enormous amount of energy in the form of light and heat, which radiates outward from the Sun and provides energy for life on Earth.

The Sun's magnetic field is also a major factor in its activity, and it drives phenomena such as sunspots, solar flares, and coronal mass ejections. These can have significant effects on the Earth's magnetic field and atmosphere and can cause disruptions to communication and power systems.

The Sun has several distinct layers, including the core, radiative zone, and convective zone, as well as the outer atmosphere, which includes the chromosphere, transition region, and corona. Each of these layers has its unique properties and behaviors, and studying them is important for understanding the Sun's activity and its effects on our Solar System.

The Sun has been studied for thousands of years and is a key focus of modern astronomy and space science. Researchers use a variety of instruments and spacecraft to study the Sun's activity, including ground-based telescopes, solar observatories, and spacecraft such as the Solar and Heliosphere

Observatory (SOHO) and the Parker Solar Probe. Understanding the Sun and its behavior is critical for predicting and mitigating the potential effects of space weather on our technology and infrastructure.

Solar flare

A solar flare is a sudden and intense burst of energy that is released from the surface of the Sun. Solar flares are caused by the sudden release of magnetic energy that has built up in the Sun's atmosphere, typically in regions called active regions that are associated with sunspots.

During a solar flare, a large amount of energy is released in the form of electromagnetic radiation, including X-rays and gamma rays, as well as energetic particles such as protons and electrons. These high-energy particles can be accelerated to very high speeds and can travel through space, potentially reaching the Earth and affecting our planet's magnetic field and ionosphere.

Solar flares can be very powerful and can release energy equivalent to billions of atomic bombs. They can cause disruptions to radio communications, GPS systems, and power grids, and can also pose a risk to astronauts and spacecraft in orbit.

Scientists study solar flares to better understand the processes that drive the Sun's activity and to better predict and mitigate the potential impacts of solar activity on our technology and infrastructure.

Solar eclipse

A solar eclipse is a phenomenon that occurs when the Moon passes between the Earth and the Sun, blocking the Sun's light and casting a shadow on the Earth. During a total solar eclipse, the Moon completely blocks the Sun's disk, allowing the Sun's outer atmosphere, called the corona, to be seen as a bright ring around the Moon.

Solar eclipses are rare events that occur only when the Sun, Moon, and Earth are perfectly aligned. Because the Moon's orbit around the Earth is tilted concerning the Earth's orbit around the Sun, solar eclipses do not occur every month, but rather at specific times during the year.

A total solar eclipse can only be seen from a narrow path on the Earth's surface, called the path of totality, which typically ranges from 50 to 100 miles wide. Observers outside this path will see a partial solar eclipse, where only a portion of the Sun is blocked by the Moon.

Solar eclipses are amazing natural spectacles and have fascinated people for thousands of years. They are important events for astronomers, who use them to study the Sun's corona and the effects of its radiation on the Earth's atmosphere. Solar eclipses can also have cultural significance in many societies and have been the subject of myth and legend throughout history.

However, it is important to remember that looking directly at the

Sun, even during a solar eclipse, can cause serious eye damage or blindness. Special filters or eclipse glasses should always be used to observe a solar eclipse safely.

Aurora

An aurora is a natural light display in the sky that occurs when charged particles from the Sun, known as the solar wind, interact with the Earth's magnetic field and atmosphere. Auroras can appear in many different colors but are most commonly seen as shimmering curtains or bands of light that glow green, red, blue, or purple.

Auroras are most commonly seen in the polar regions, particularly in the high latitudes of the northern and southern hemispheres. The phenomenon is known as the aurora borealis, or northern lights, in the north, and the aurora australis, or southern lights, in the south.

The Earth's magnetic field channels the charged particles from the solar wind toward the poles, where they collide with gas molecules in the Earth's upper atmosphere. This collision causes the gas molecules to become excited and emits light, producing the bright and colorful displays of an aurora.

Auroras are influenced by many factors, including solar activity, the Earth's magnetic field, and the composition of the Earth's upper atmosphere. They are often seen during periods of increased solar activity, such as during a solar storm or a coronal mass ejection, which can send large amounts of charged particles toward the Earth.

Auroras are a stunning natural phenomenon that has fascinated people for centuries. They are also important for scientific research, as they provide insight into the behavior of the Sun and the Earth's magnetic field, and can help us better understand the complex interactions between our planet and the space environment.

Other Celestial Body
Dwarf planet

A dwarf planet is a type of celestial body that orbits the sun but is not considered a full-fledged planet. This designation was created in 2006 by the International Astronomical Union (IAU) in response to the discovery of several objects in our solar system that was previously classified as planets but did not meet the criteria for full planetary status.

To be classified as a dwarf planet, a celestial body must meet three criteria:

- It must orbit the sun.
- It must have sufficient mass to assume a nearly round shape.
- It must have cleared its orbit of other debris.

The last criterion is the most significant difference between planets and dwarf planets. Planets have cleared their orbits of other debris, whereas dwarf planets have not. This means that they share their orbits with other bodies, such as asteroids or other dwarf planets.

There are currently five officially recognized dwarf planets in our solar system: Pluto, Ceres, Haumea, Makemake, and Eris. All of these objects are located beyond the orbit of Neptune in a region known as the Kuiper Belt, except for Ceres, which is located in the asteroid belt between Mars and Jupiter.

Asteroid

An asteroid is a small, rocky object that orbits the sun. Asteroids are sometimes called minor planets because they are too small to be considered planets, but they share many similarities with planets in terms of their composition and orbit.

Most asteroids are found in the asteroid belt, a region located between the orbits of Mars and Jupiter. This region contains millions of asteroids, ranging in size from a few meters to hundreds of kilometers in diameter. Some asteroids have irregular shapes, while others are nearly spherical.

Asteroids are believed to be remnants from the early solar system, leftover from the formation of the planets. They are composed of rock, metal, and sometimes ice, and their composition can vary widely from one asteroid to another. Some asteroids are also believed to contain valuable resources, such as metals and water, which could be mined in the future for use in space exploration.

Asteroids can pose a potential threat to Earth if they collide with our planet. While the probability of a major asteroid impact is low, efforts are underway to detect and track potentially hazardous asteroids, as well as to develop technologies to deflect them if necessary.

Comets

Comets are small, icy celestial bodies that orbit the Sun. They are made up of a mixture of frozen water, gases, dust, and rock, and are sometimes referred to as "dirty snowballs" because of their composition.

When a comet's orbit brings it close to the Sun, the heat causes the frozen gases and dust to vaporize, creating a glowing coma (a cloud of gas and dust) around the comet's nucleus. This can sometimes be visible from Earth as a bright tail extending away from the Sun. The tail always points away from the Sun due to the solar wind pushing the gas and dust particles away.

Comets can have highly elliptical orbits that take them far out into the solar system and then back in toward the Sun. Some comets have extremely long orbital periods, taking hundreds or thousands of years to complete a single orbit.

Comets are believed to be remnants from the formation of the solar system, and studying them can provide important insights into the early stages of our solar system's history. Several spacecraft have been sent to study comets up close, including the European Space Agency's Rosetta mission, which successfully landed a probe on the surface of a comet in 2014

Metroid

A meteoroid is a small piece of rock or debris in space that is smaller than an asteroid but larger than a molecule. Meteoroids range in size from tiny dust particles to objects several meters in diameter. They are usually the remnants of comets or asteroids that have been broken apart through collisions or other processes.

When a meteoroid enters Earth's atmosphere, it is referred to as a meteor or shooting star. The intense heat generated by friction with the atmosphere causes the meteoroid to burn up, creating a bright streak of light in the sky. Most meteors are very small and burn up completely before reaching the ground, but larger meteoroids can survive the journey and impact the Earth's surface.

Meteoroids are also known to be responsible for creating impact craters on the surfaces of planets and moons in our solar system. The largest meteorite impact on Earth is believed to have occurred in modern-day South Africa around 2 billion years ago and created the Vredefort crater, which is estimated to be 300 kilometers (186 miles) in diameter.

Meteoroids and their associated phenomena are studied by

astronomers and planetary scientists to better understand the composition and history of our solar system

Kuiper belt object (KBO)

A Kuiper Belt Object (KBO) is a type of small, icy object that orbits the sun in a region beyond the orbit of Neptune, known as the Kuiper Belt. This region is similar to the asteroid belt, which is located between the orbits of Mars and Jupiter, but contains a much larger number of objects and is dominated by icy bodies rather than rocky ones.

KBOs are believed to be remnants from the early solar system, formed from the same material that created the outer planets. They are typically composed of water, methane, ammonia, and other ices, as well as rocky materials. Some KBOs are large enough to be considered dwarf planets, such as Pluto and Eris.

The Kuiper Belt is also the source of many short-period comets, which are believed to be KBOs that have been gravitationally perturbed by the gas giants and sent into the inner solar system. Studying KBOs can therefore provide insights into the early history and evolution of the solar system.

KBOs are difficult to observe and study due to their distance from Earth and their small size. However, several spacecraft missions have been launched to study KBOs up close, including the New Horizons mission, which flew by Pluto in 2015 and is now on route to study other KBOs in the Kuiper Belt.

Exoplanet

An exoplanet, or extrasolar planet, is a planet that orbits a star outside our solar system. Since the first exoplanet was discovered in 1995, thousands of others have been identified by astronomers using a variety of techniques.

Exoplanets come in a wide range of sizes and compositions, from small, rocky planets like Earth to giant gas planets like Jupiter. Some exoplanets are located within the habitable zone of their star, where the temperature is just right to allow liquid water to exist on the planet's surface, making them potentially habitable for life as we know it.

Exoplanets are discovered through a variety of methods, including the transit method, in which astronomers measure the slight dip in a star's brightness when a planet passes in front of it, and the radial velocity method, in which astronomers detect the wobble of a star caused by the gravitational pull of an orbiting planet.

The study of exoplanets has revolutionized our understanding of the universe and our place in it. It has provided evidence that there are many other planets beyond our solar system and has expanded our knowledge of planetary systems and their evolution. The search for exoplanets continues to be an active area of research in astronomy and astrophysics

Nebulae -

Nebulae are vast clouds of gas and dust in space, often illuminated by the light of nearby stars. They are some of the most visually stunning objects in the universe and are important sites of star formation.

Nebulae can be classified into different types based on their appearance and composition. Some of the most common types include:

1. *Emission nebulae*: These nebulae are primarily made up of ionized gas that emits light of various colors, often appearing red or pink. They are often the sites of active star formation and are illuminated by young, hot stars.

2. *Reflection nebulae*: These nebulae are made up of dust particles that reflect the light of nearby stars. They often appear

blue and are sometimes seen surrounding emission nebulae.

3. *Dark nebulae*: These are dense clouds of gas and dust that block the light of stars behind them, appearing as dark patches against the background sky. They are often the sites of future star formation.

4. *Planetary nebulae*: These nebulae are formed by the outer layers of dying stars that have been ejected into space. They often have a characteristic spherical or disk-like shape and are illuminated by the central star, which has become a white dwarf.

Nebulae are important sites for studying the processes of star formation and the chemical evolution of the universe. They are also some of the most beautiful and awe-inspiring objects in the sky, visible with telescopes and even with the naked eye under dark skies.

Galaxies

Galaxies are massive, gravitationally bound systems of stars, dust, gas, and dark matter that are distributed throughout the universe. There are billions of galaxies in the observable universe, ranging in size from small, dwarf galaxies to massive, giant elliptical galaxies.

The three main types of galaxies are spiral, elliptical, and irregular. Spiral galaxies have a distinct spiral shape with a central bulge surrounded by a disk of stars, dust, and gas. Elliptical galaxies are more rounded in shape and have a smooth distribution of stars. Irregular galaxies have a more chaotic shape and no clear structure.

Galaxies are believed to have formed from the gravitational collapse of clouds of gas and dust in the early universe. Over time, galaxies can merge, creating even larger and more complex structures.

The study of galaxies is a major area of research in astronomy and astrophysics. Astronomers use a variety of telescopes and observational techniques to study the properties of galaxies, including their size, shape, composition, and movement. Understanding the evolution and properties of galaxies can provide important insights into the history and structure of the universe as a whole.

Black hole

A black hole is an object in space with a such strong gravitational pull that nothing, not even light, can escape its grasp. Black holes are formed when massive stars' collapse under their own gravity at the end of their lives, creating a region of space with infinite density and zero volume known as a singularity.

Black holes are invisible, as they do not emit any light or other forms of radiation that we can detect. However, their presence can be inferred by the effects of their gravity on nearby matter, such as stars and gas clouds.

There are three main types of black holes: stellar black holes, intermediate black holes, and supermassive black holes. Stellar black holes are formed from the collapse of individual massive stars and can have a mass up to *20* times that of the sun. Intermediate black holes are believed to form from the merging of smaller black holes and can have masses between 100 and *100,000* times that of the sun. Supermassive black holes, on the other hand, are thought to reside at the centers of most galaxies, including our own Milky Way, and can have masses up to billions of times that of the sun.

Black holes are fascinating objects that challenge our understanding of the universe and the laws of physics. They have been the subject of intense study by astronomers and physicists, and their discovery has led to new insights into the nature of space, time, and gravity.

Neutron Star

A neutron star is a type of extremely dense and compact star that is formed from the remnants of a massive star that has undergone a supernova explosion. They are incredibly dense, with a mass similar to that of the sun packed into a region of space only a few kilometers in diameter.

The intense gravitational forces present in a neutron star cause its atoms to break down, leaving only a dense soup of neutrons. This gives neutron stars their name. They are so dense that a single teaspoon of neutron star material would weigh several billion tons on Earth.

Neutron stars have extremely strong magnetic fields, and they emit beams of radiation that can be observed as pulsars. Pulsars are rapidly rotating neutron stars that emit regular pulses of radiation as they spin. These pulses can be detected with radio telescopes and have been used for a variety of scientific studies, including tests of general relativity and the study of the interstellar medium.

Neutron stars can also form in binary systems, where they orbit around a companion star. In some cases, material from the companion star can be accreted onto the neutron star, leading to X-ray emission.

Neutron stars are important objects in astrophysics and are studied across a wide range of wavelengths, from radio waves to X-rays and gamma rays. They are also believed to be the progenitors of some types of supernova explosions, and their study has led to important insights into the physics of extreme states of matter and the behavior of matter in strong gravitational fields.

Chapter 2

Satellites

A satellite is an object that orbits another object in space. Satellites can be natural, such as the moon orbiting the Earth, or artificial, such as spacecraft launched by humans.

Artificial satellites are used for a wide range of applications, including:

- *Communications*: Satellites can be used to transmit signals for television, radio, and other communication services. They can also be used to provide internet access to remote areas.
- Navigation: Global Navigation Satellite Systems (GNSS), such as the Global Positioning System (GPS), use satellites to provide

accurate positioning and timing information for navigation.

- Earth observation: Satellites equipped with sensors and cameras can be used to study the Earth's environment, weather patterns, and natural disasters. They can also be used for monitoring crops, forests, and other land use patterns.

- Science: Satellites can be used for a wide range of scientific research, including studying the universe, observing other planets and their moons, and monitoring the sun and its effects on Earth.

- *Defense:* Satellites can be used for military and defense purposes, including intelligence gathering and surveillance.

Satellites can be placed in different orbits depending on their intended use. Low Earth Orbit (LEO) is used for Earth observation and communication, while Geostationary Orbit (GEO) is used for communication and navigation. Polar orbits are used for Earth observation and scientific research.

The use of satellites has revolutionized many aspects of our lives, from communication and navigation to weather forecasting and disaster management. Satellites will continue to play a vital role in our technological infrastructure and scientific research in the years to come.

Here are some additional facts about satellites:

1. There are thousands of artificial satellites currently orbiting the Earth, with various purposes and missions.
2. Satellites can be placed in different orbits depending on their intended use. Some of the most common orbits include Low Earth Orbit (LEO), Medium Earth Orbit (MEO), Geostationary Orbit (GEO), and Highly Elliptical Orbit (HEO).

3. Satellites can be powered by solar panels or batteries, depending on their mission and expected duration in space.
4. The International Space Station (ISS) is a habitable artificial satellite that orbits the Earth at an altitude of approximately 408 kilometers (253 miles). It is a joint project of five space agencies: NASA (United States), Roscosmos (Russia), JAXA (Japan), ESA (Europe), and CSA (Canada).
5. Satellites can experience various challenges and risks, such as space debris, radiation, and micrometeoroids, which can

cause damage or even failure of the satellite.

6. Satellites can be launched into orbit using different methods, including rockets, space planes, and balloons.

7. The cost of building, launching, and maintaining a satellite can vary widely depending on its size, mission, and other factors.

8. In addition to Earth orbiting satellites, there are also satellites that have been sent to explore other planets and moons in our solar system, such as the Mars Reconnaissance Orbiter and the Cassini spacecraft, which explored Saturn and its moons.

9. The use of small satellites, known as CubeSat's, has become increasingly popular in recent years, as they can be more affordable and easier to launch than larger satellites.

Overall, satellites have revolutionized many aspects of our lives and have become an essential part of our technological infrastructure

A space station is a large spacecraft designed to be habitable for humans, orbiting the Earth or other celestial bodies. The primary purpose of a space station is to provide a long-term base for scientific research and experimentation in space, as well as to support space exploration and human spaceflight missions.

The most well-known space station is the International Space Station (ISS), a joint project of five space agencies: NASA (United States), Roscosmos (Russia), JAXA (Japan), ESA (Europe), and CSA (Canada). The ISS orbits the Earth at an altitude of approximately 408 kilometers (253 miles) and has been continuously occupied since November 2000. It is the largest artificial body in orbit and is roughly the size of a football field.

The ISS serves as a platform for a wide range of scientific research, including biology, physics, and materials science. It also supports experiments in astronomy, meteorology, and Earth observation. In addition, the ISS is used to test technologies and techniques for future human space exploration missions, such as life support systems, propulsion, and radiation shielding.

The ISS is equipped with various modules, or sections, that provide living quarters for the crew, as well as laboratories, storage areas, and docking ports for visiting spacecraft. The station is powered by solar panels, which convert sunlight into electrical power, and is supplied with food, water, and other essentials by regular resupply missions from Earth.

In addition to the ISS, there have been other space stations throughout history, including the Soviet/Russian Mir space station and the Chinese Tianlong space station. Future space stations are also planned, including the Lunar Gateway, a planned space station in lunar orbit that will serve as a base for future human missions to the moon and beyond.

Overall, space stations play a vital role in advancing scientific knowledge and expanding our capabilities for space exploration. They also serve as a symbol of international cooperation and collaboration in space exploration.

James webb

The *James Webb Space Telescope (JWST)* is a large, infrared telescope that is set to launch in December 2021. It is named after James E. Webb, who served as the second administrator of NASA from 1961 to 1968 and played a key role in the development of the Apollo program.

The JWST is designed to be the successor to the Hubble Space Telescope, with the ability to observe the universe in greater detail and depth than ever before. It will be positioned at a distance of about 1.5 million kilometers (930,000 miles) from Earth, at a location known as the second Lagrange point (L2).

The JWST has a large, segmented primary mirror that is 6.5 meters (*21.3 feet*) in diameter, which is more than two and a half times the size of Hubble's mirror. It is also equipped with advanced scientific instruments that will allow it to observe the universe in the infrared spectrum, which can penetrate dust clouds and reveal details about the formation and evolution of galaxies, stars, and planetary systems.

The JWST has been in development for over 20 years, with contributions from NASA, the European Space Agency (*ESA)*, and the Canadian Space Agency (*CSA*). It has faced numerous technical and budgetary challenges over the years but is now on track for its long-awaited launch.

The JWST is expected to revolutionize our understanding of the universe and help answer some of the biggest questions in astrophysics, such as the origins of the first galaxies, the formation of stars and planets, and the search for signs of life on other planets.

Hubble telescope

The Hubble Space Telescope is a large, space-based observatory that was launched into orbit around the Earth in 1990. It was named after the American astronomer Edwin Hubble, who is known for his groundbreaking research on the expansion of the universe.

The Hubble telescope is a joint project of NASA and the European Space Agency (*ESA*), and it has been one of the most important tools in modern astronomy. It is equipped with a large, 2.4-meter (*7.9-foot*) diameter mirror that collects visible, ultraviolet, and near-infrared light from distant objects in space. It orbits the Earth at an altitude of about 540 kilometers (*335 miles*) and circles the planet about once every 97 minutes.

The Hubble telescope has made many important discoveries and observations over the years, including:

- The first accurate measurement of the expansion rate of the universe, which helped to confirm the Big Bang theory of the origin of the universe.
- The discovery of black holes in the centers of galaxies, has led to a better understanding of the structure and evolution of galaxies.
- The detection of dark energy, is a mysterious force that is causing the expansion of the universe to accelerate.
- The identification of thousands of new galaxies, including some

of the most distant and oldest known objects in the universe.

- The study of planetary systems outside our own Solar System, including the discovery of new exoplanets and the analysis of their atmospheres.

The Hubble telescope has been serviced and upgraded several times over the years by space shuttle missions, which have replaced old components and installed new scientific instruments. It is expected to remain in operation until at least the mid-2020s when it will be replaced by the James Webb Space Telescope.

Chapter 3

All Recent Discoveries in Space

Here are some recent discoveries in space that were made prior to that time:

1. *Water on the Moon*: In October 2020, NASA announced that its Stratospheric Observatory for Infrared Astronomy (SOFIA) had detected water molecules on the sunlit surface of the Moon. This discovery suggests that water may be more abundant on the Moon than previously

thought, which could be important for future lunar exploration.

2. *Phosphine on Venus*: In September 2020, a team of scientists announced that they had detected phosphine gas in the atmosphere of Venus. This discovery was surprising because phosphine is a gas that is typically associated with life on Earth. However, the scientists cautioned that the presence of phosphine alone does not necessarily mean that there is life on Venus.

3. *Fast Radio Bursts*: Fast Radio Bursts (FRBs) are brief, intense

bursts of radio waves that come from distant galaxies. In 2020, astronomers announced the discovery of a new repeating FRB, which allowed them to study the source in more detail. They also detected a new non-repeating FRB that was particularly strong, which could help to shed light on the mysterious origins of these phenomena.

4. *Black Hole Collision*: In September 2019, the Laser Interferometer Gravitational-Wave Observatory (LIGO) and the Virgo Collaboration announced the detection of a

collision between two black holes. This was the first black hole collision ever detected that produced a gravitational wave signal in all three of these observatories.

5. *First Image of a Black Hole*: In April 2019, the Event Horizon Telescope (EHT) collaboration announced that they had captured the first-ever image of a black hole. The image showed the supermassive black hole at the center of the galaxy Messier 87, which is located about 55 million light-years away from Earth. This was a major milestone in the study of black

holes and the structure of the universe.

6. *Gravitational Waves*: In 2015, scientists made the first detection of gravitational waves, which are ripples in the fabric of space-time caused by the acceleration of massive objects. This discovery confirmed a key prediction of Einstein's theory of general relativity and opened up a new way of studying the universe.

Some other discoveries

- The discovery of the Higgs boson (2012): The Higgs boson, a particle that gives other particles mass, was discovered in 2012 at the Large Hadron Collider.
- The discovery of the interstellar medium (1930s): In the 1930s, astronomers discovered that the space between stars is not empty, but filled with gas and dust known as the interstellar medium.
- The discovery of dark matter (1930s-1970s): In the 1930s, astronomers discovered that there is more mass in the

universe than can be accounted for by visible matter, leading to the discovery of dark matter.

- The discovery of cosmic microwave background radiation (1965): In 1965, scientists discovered cosmic microwave background radiation, the faint afterglow of the Big Bang that fills the entire universe.

- The discovery of the Kuiper Belt (1992): The Kuiper Belt, a region of the solar system beyond the orbit of Neptune that is home to numerous icy objects, was discovered in 1992.

- The discovery of exomoons (2018): In 2018, the first exomoon, a moon orbiting an exoplanet, was discovered.
- The discovery of the first black hole image (2019): In 2019, scientists released the first-ever image of a black hole, captured using a network of radio telescopes known as the Event Horizon Telescope.

As you close the final pages of this book on space technology, take a moment to reflect on the incredible advancements and achievements that have been made in this field. From the first human journey into space to the development of reusable rockets, from the exploration of our solar system to the discovery of exoplanets, space technology has expanded our understanding of the universe and transformed our world in countless ways.

But this is only the beginning. As we continue to push the boundaries of what is possible, we will unlock new technologies and capabilities that will enable us to explore even further and accomplish even more. From establishing a sustainable human presence on the Moon and Mars to discovering new forms of life beyond our planet, the future of space technology is full of endless possibilities.

So, whether you are a student, a researcher, an engineer, or simply someone with a passion for space exploration, remember that you too can contribute to this exciting and ever-evolving field. With dedication, creativity, and perseverance, we can continue to push the limits of what we know and discover what lies beyond. The universe is waiting for us, and the future of space technology is in our hands.

The End